A BIG YELLOW BOOK

Giant Dump Trucks

Written by Jean Eick
Illustrated by Michael Sellner

VENTURE
PUBLISHING
St. Paul, MN

A BIG YELLOW BOOK

Published by Venture Publishing
315 5th Ave. NW, St. Paul, MN 55112

Designed by Michael Sellner
Edited by Jackie Taylor
Production: James Tower Media • Design
Photo Credits: "Images © 1995 PhotoDisc, Inc." Pages: 22, 23
All other photos courtesy of Caterpillar Image Lab

Printed in the United States of America

Library of Congress Cataloging-in-Publication Data
Eick, Jean, 1947-
    Giant dump trucks / written by Jean Eick; illustrated by Michael Sellner.
        p.   cm. – (A big yellow book)
Summary: Presents facts about dump trucks, their working parts, what they
are used for, and why they are important.
        ISBN 1-888637-07-2
        1. Dump trucks – Juvenile literature.   [1. Dump trucks. 2. Trucks.]
I. Sellner, Michael, ill. II. Title. III. Series.
TL230.15.E33   1996
629.225—dc20   96-4096
                                                                CIP
                                                                 AC

Contents

What is a Giant Dump Truck?   4-7

Parts of a Giant Dump Truck   8-9

Inside the Cab   10-11

Uses for Giant Dump Trucks   12-19

Special Facts About Giant Dump Trucks   20-21

Where in the World Can You Find
Giant Dump Trucks?   22-29

Putting It All Together   30-31

Words to Remember   32

What is a Giant Dump Truck?

Have you ever seen a dump truck tilt up its back and dump a load of rocks or dirt? Dump trucks are specially built for moving heavy loads from one place to another. They carry the load and use arms, called hoists, to lift the body and dump the load.

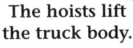
The hoists lift
the truck body.

People use regular dump trucks to build roads and do other important jobs, but some extra large jobs need extra large trucks. Giant dump trucks do the jobs that are too big for regular dump trucks.

Imagine a dump truck the size of a house, roaring along as dust flies all around it. You won't see these giant dump trucks traveling down the road. That's because they work with other big machines in mines, quarries and other jobs off the highway.

Giant dump trucks hard at work in a quarry and an open pit mine.

Giant dump trucks are built to carry enormous loads.
They are powerful machines!

Parts of a Giant Dump Truck

Canopy - A hard covering over the cab that protects the operator.

Mirrors - Located on both sides of the truck to help the operator see in all directions.

Stairs - Used by the operator to get into the cab.

Engine - Where the power comes from to run the machine.

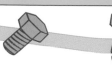

Cab - The area where the operator sits to run the truck.

Body - The area on the back of the truck that carries the load.

Tires - Four or six wheels made of hard rubber that allow the truck to move.

Inside the Cab

The place where the operator sits is called the cab. Giant dump trucks are so tall that the operator has to climb up the stairs to get into it. The cab is like a little room, with a door and windows all around it.

Climbing up to the cab.

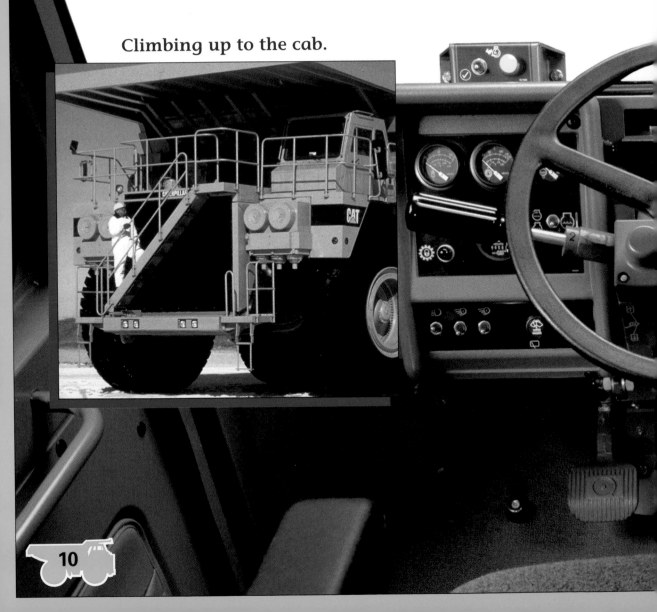

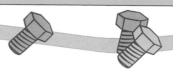

Giant dump trucks sometimes work in dangerous places. The special canopy over the cab protects the operator from anything that falls on top. The operator sits in a big seat with extra cushions. All of the pedals and levers are easy to reach. Special lights inside let the operator know whether the truck is running properly or the load is full.

Window

Steering Wheel

Levers

Pedals

Seat

Uses for Giant Dump Trucks

Giant dump trucks are made for working in mines and quarries. Open pit mines are used to mine gold, copper, coal and iron. Mines can be very deep in the ground. Trucks must be able to go up and down very steep roads to haul rocks out of these mines.

Coming back for another load.

Rocks and minerals are also found in quarries. A quarry is not as deep as a mine, but it is still a place for hard working giant dump trucks.

Trucks at work in a quarry.

Trucks at work in an open pit mine.

Mining and quarry trucks usually work with machines that have giant front buckets like excavators and wheel loaders. A wheel loader can fill the body of the truck in about seven dumps.

A gigantic front shovel fills a truck in an open pit mine.

The quarry truck has a flat body that allows heavy loads to slide out easily. The operator can dump the load into a machine that will crush the rocks.

Quarry trucks haul very heavy loads.

The load slides right down the flat body.

In some open pit coal mines, giant machines remove the coal from the walls and push it into piles on the bottom of the pit.

A wheel loader scoops up the coal and puts it into the body of a giant dump truck. Many of these mines are so big they need an entire fleet of giant dump trucks!

Sometimes a giant dump truck has to turn in very tight spaces. Articulated trucks are built to bend in the middle. They use a special link, called a hitch, to attach the body to the cab. The operator can turn the body to the right or left, for easy loading.

The hitch allows this truck to make sharp turns.

Articulated trucks don't travel as far as other trucks, but they are easier to use in tight spaces.

Special Facts About
Giant Dump Trucks

Unless you have stood next to one of these giant dump trucks, it's hard to imagine just how big they really are. Giant dump trucks come in different sizes. The smallest models are very big, but they don't seem that way next to the largest models. That's because those trucks are enormous!

🔩 How Tall Are Giant Dump Trucks?

When the smallest of these giant trucks tilts up to dump its load, it is as tall as a two story building (16 feet 5 inches/5.00 meters).

When the largest one tilts up, it is as tall as a four story building (43 feet 4 inches/13.21 meters).

Largest Dump Truck

🔩 How Much Fuel Do Giant Dump Trucks Hold?

The smallest model holds enough fuel for about three cars (55 gallons/210 liters).

You could fill up 50 cars with the fuel from just one of the large trucks (1,020 gallons/3,861 liters).

🔩 How Much Do Giant Dump Trucks Weigh?

The smallest model weighs more than two full grown elephants (33,070 pounds/15,000 kilograms). That's a heavy dump truck!

It takes ten smaller trucks to weigh as much as just one large dump truck (323,709 pounds/143,564 kilograms). That's more than 20 elephants!

🔩 How Fast Can A Giant Dump Truck Go?

When giant dump trucks have a full load, they can go about as fast as cars travel on neighborhood streets.

The smallest model can go 28.6 miles per hour (46 kilometers per hour).
The largest model can go 34 miles per hour (54.7 kilometers per hour).

Smallest Dump Truck

Where in the world can you find a Giant Dump Truck?

Giant dump trucks are found all over the world. In most countries they are found working in mines and quarries.

In Scotland, they haul granite. In Australia and West Africa, they haul coal and gold. In Northern Siberia, they are used in diamond mines. In France, giant dump trucks are used on special road projects that are too big for regular dump trucks.

North America

South America

Scotland

Northern
Siberia

Europe

Scotland

France

Northern
Siberia

Asia

West
Africa

Australia

Africa

Australia

Antarctica

France

Australia

West Africa

23

Where in the world can you find a Giant Dump Truck?

Thailand

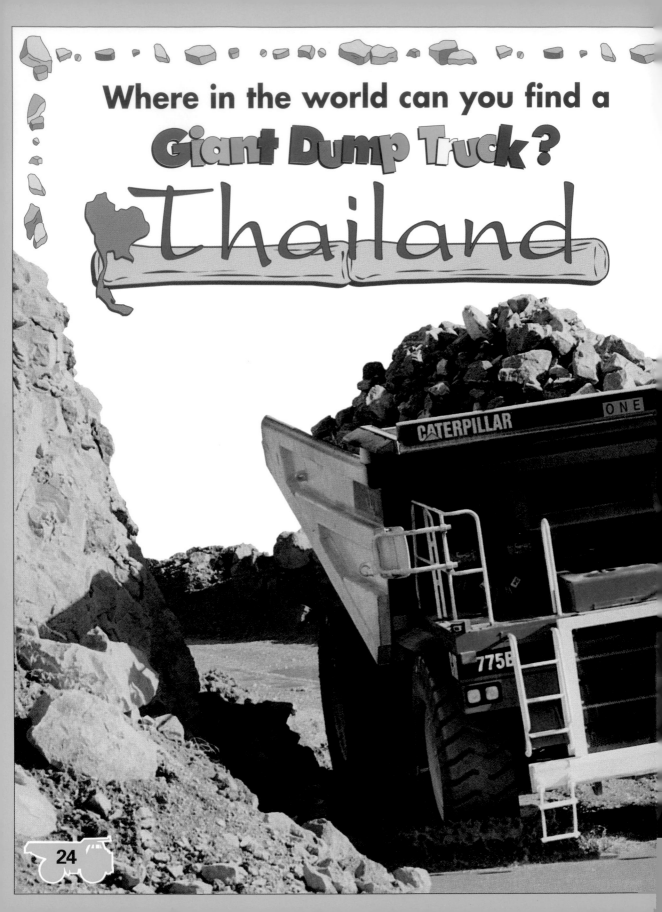

CATERPILLAR   ONE

775B

ining is an important business in Thailand. Many giant dump trucks are needed to work in large open pit coal mines, cement plants and rock quarries.

QUARRY TRUCK

75B

Where in the world can you find a Giant Dump Truck?
United States

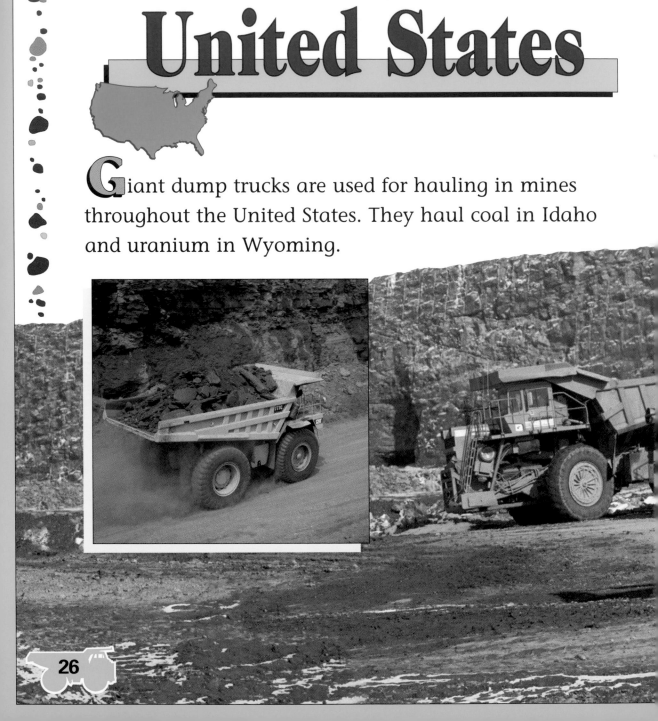

Giant dump trucks are used for hauling in mines throughout the United States. They haul coal in Idaho and uranium in Wyoming.

These trucks are also needed to haul limestone and other rocks out of quarries in states like Virginia and Minnesota.

Excavators and wheel loaders work with the giant dump trucks in this open pit mine.

27

Where in the world can you find a Giant Dump Truck?

Colombia

The world's largest open pit coal mine is located in northern Colombia. Giant dump trucks work extra hard hauling coal from the bottom of this pit.

Where in the world can you find a
Giant Dump Truck?

Giant dump trucks help people all over the world. Some of the huge loads they haul provide us with fuel, gold and diamonds. Giant dump trucks are big, powerful machines that help shape our world.

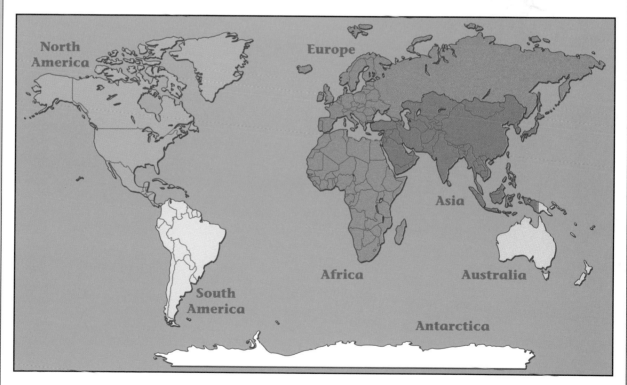

North America

Europe

Asia

Africa

Australia

South America

Antarctica

Putting It All Together

Putting a giant dump truck together is a huge job. Many men and women work together to make the individual parts of the trucks. Most giant dump trucks are too big to be built in a factory, so they are put together at the job site.

Even the tires look enormous.

This newly assembled
truck is ready to work.

Words To Remember

Articulated Truck - Connected by a special link, called a hitch, allowing the truck to bend in the middle.

Engine - Where the power comes from to run the machine.

Giant dump truck - One of the huge trucks used for mining.

Open pit mine - A deep area cut into the earth's surface with steep roads going in and out of the pit.

Quarry - An open area cut into the ground for mining rocks.